Photos: B. Bates, A. Bates, S. Flannery
Art: A. J. Bates
Little Pond Stories, LLC

ISBN: 978-1-7373520-1-3

Library of Congress: 2021912165

Dedication

Dear nature,
thank you for loving us.

Hello,
fellow nature
friends!
Are you ready for
an adventure?

Listen closely to a story about a place I call Little Pond. I hope you can find a pond of your own one day!

When you do, please make sure to love the plants and animals. They are an important part of the world, just like you.

Where did I find a pond of my own?

I discovered Little Pond after I found a map in my attic! It was dusty and covered with old spider webs. Old things tell us stories, much like places.

On the map, there is a magical spot between the long grass and an old forest. The water is filled with fish, frogs, turtles, and more! There are birds in the trees and bugs in the reeds.

Years ago, Mr. and Mrs. Beaver built a dam of sticks, mud, and stones to hold back the water.

Their house, called a **lodge**, grows a bit larger year after year. The beavers use the pond water as protection from **predators**. Some of these are coyotes, bears, foxes, and bobcats.

Thanks to the beavers, all of the animals have **shelter** and enjoy the water. Do you see the goose?

Many plants grow near and in the water. These feed the caterpillars that then turn into butterflies! The flowers feed bees all spring, summer, and fall. Insects love Little Pond.

Singing birds, perching birds, and **waterfowl** make their homes here. The pond and shore are places to explore for kingfishers, herons, mallards, and more!

But life at Little Pond was not always this way.

Before the beavers built the dam, Little Pond was not a pond at all. It was a sad and squiggly brook. When the beavers came to live here, everything changed.

Once the dam was built, the water began to pool behind its wall. This created Little Pond, a new home for all animals who live in a special **habitat**.

Beavers work hard to keep the dam safe. They add or remove stones and sticks to keep the water as deep as they need. Can you see Mrs. Beaver?

But one spring, the beavers did not come out. Using my notebook to keep **observations** of what happens at the pond, I noticed there was a problem.

Spring came and went. Summer began, and Mallard and I observed something strange at the dam.

The dam was leaking! First, there were tiny trickles, Then, small holes opened up. More and more water drained out.

Little Pond was shrinking. Without the water and with the heat of the summer sun, the habitat for fish, frogs, turtles, and waterfowl was in danger.

When I visit a magical place, I carry a science book and make **observations** of the **environment**. I write down and draw what I see, hear, smell, and even taste in the air.

I began my **investigation** by asking a question.

Why did the beavers leave Little Pond?

When I observe, I take notes and listen to what is going on around me. I began by looking at the pond, field, and forest for clues.

Suddenly, I heard a rattle in the sky! King, the Belted Kingfisher, is a waterfowl. He builds his home along the shore. He digs into the **bank** of dirt to make a **burrow**.

He landed on a branch, but kept secret. Below him, tiny creek chubs and crayfish swam out of sight.

"Wow! Look at this!" I said out loud to King. I bent down to see a snapping turtle in the grass. She was on her way to dig a hole in the ground and lay eggs.

"Hello, Snapper," I whispered. "I will stay here, and you can stay over there. I know from my reading that you are a shy creature and need your space."

With a great rattle, King left his perch and flew towards the forest.

I looked and saw the Pileated Woodpecker at an old stump. The big bird pecked at the ants and bugs crawling out.

I watched Pileated and then moved to the forest. On my way, I studied the shore for clues. I saw chewed tree stumps. They were dry, not new.

I saw a fence around a tree where the forest meets the shore. I decided to follow it and investigate what it was fencing in.

The birch trees had wire fencing around them. I looked from the fences, to the forest, to my farmhouse. Do you know why the beavers left?

Someone fenced the birch trees in to keep the beavers out. The beavers had no food! They lost what they needed to survive and thrive.

Beavers need wood to chew, eat, cut, and use to stay alive. To keep a dam safe and strong, beavers work with sticks and logs they **harvest** after storing food.

Beavers use every part of the trees they **fell**. They also eat greens like grasses and water plants.

The fence was **evidence** that a person tried to keep the beavers out, someone who once lived at our house. I decided to walk around a bit more.

I heard the small call of bluebirds taking baths in the water.

The young Red-Shouldered Hawk screeched from a dead branch. The dead trees around the pond have holes where birds nest.

I wondered and thought about all the birds and bugs who make their homes in them.

The sunlight shined into the water to feed the plants. When a live tree hangs over the water, it makes shade, and plants do not always grow under it.

The animals and plants living at Little Pond **survive** because of the beaver dam!

Months passed. Rains came. Water poured over and through the dam. Beavers, where are you?

Have you ever tried to make a wall to hold back water? Humans are not beavers!

Beavers are called **engineers**. They use their **instinct** to fix leaks in the dam, like putting a puzzle together.

With each day of rain, the dam was in danger of breaking apart. This is called a **failure**.

The fish, amphibians (frogs, newts, salamanders, toads), snakes, crayfish, and all **aquatic** life was **threatened**. The waterfowl would have no source of food and leave for other shores.

Think about how one person's choice to fence in trees changed an entire **ecosystem**.

But there was something helpful humans could do!

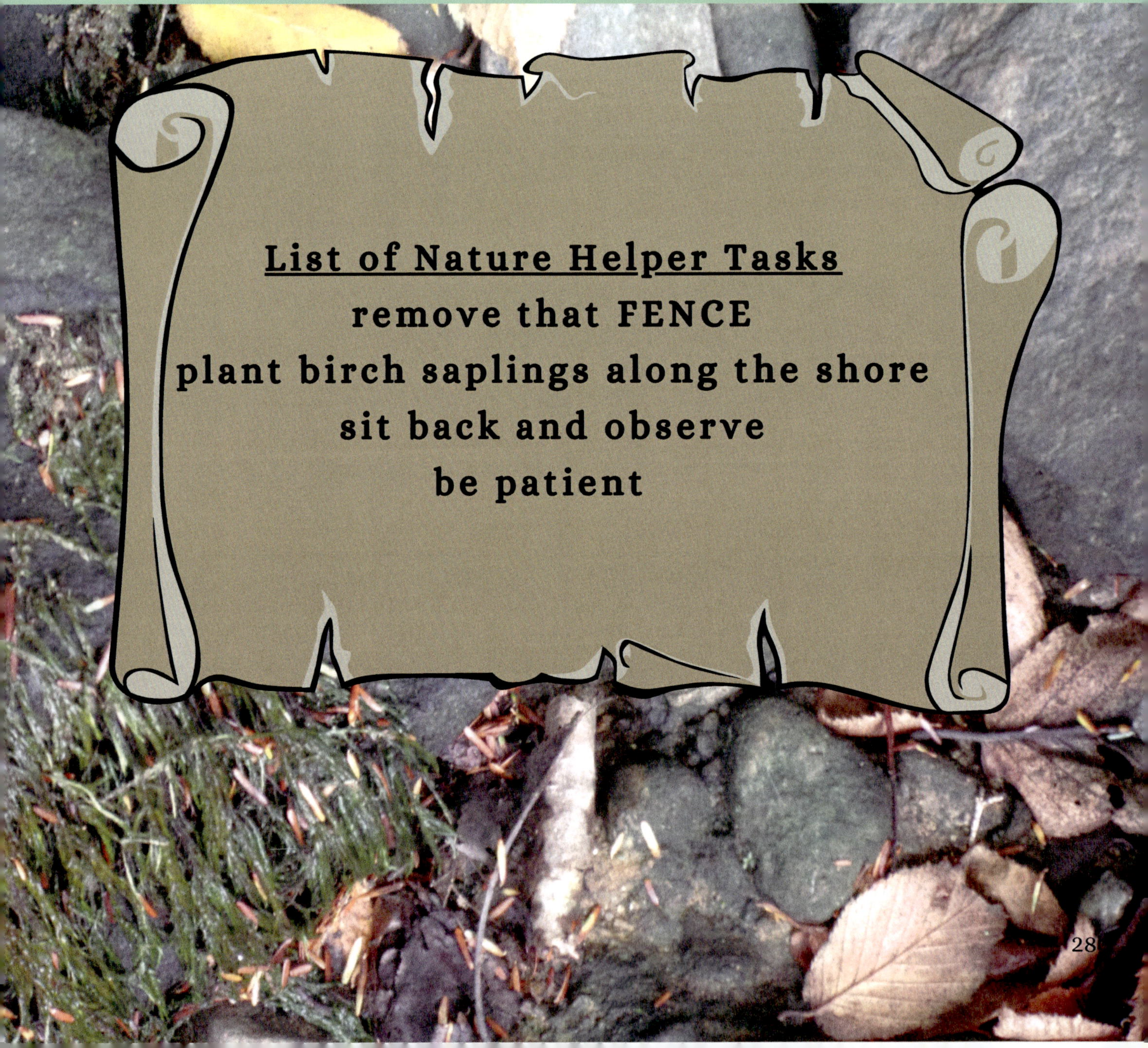

<u>List of Nature Helper Tasks</u>

remove that FENCE

plant birch saplings along the shore

sit back and observe

be patient

SPLASH!

What are those ripples in the water **this year**?

Mr. and Mrs. Beaver glide across Little Pond in front of me, using their flat tails and **webbed** feet to move. They stop and chew upon a huge branch, their orange teeth sparkling in the setting sun.

"Welcome home, my friends," I say out loud.

What a special place, this Little Pond. Promise to explore with me again soon! I am excited to see what we will find next time.

EXPLORER'S OBSERVATION JOURNAL

Words to know:

beaver dam
lodge
predator
shelter
waterfowl
habitat
observation
environment
investigation
bank (shore)
burrow

harvest
fell (trees)
evidence
survive
ecosystem
engineers
instinct
failure (dam)
aquatic
threatened
webbed

Acknowledgements

Thank you to my family, teachers, and great circle of friends for support, guidance, and advice.

And thank you to our dear beavers at Little Pond.

To learn more:

Please visit littlepondstories.org

Dear Readers,

Have nature OBSERVATIONS to share or an idea to EXPLORE? Send me an email and tell me all about it!

littlepondstories@gmail.com

Made in United States
Orlando, FL
02 November 2026

90104232R00021